LIGHT

BY SALLY M. WALKER
PHOTOGRAPHS BY ANDY KING

LERNER PUBLICATIONS COMPANY • MINNEAPOLIS

Additional photographs are reproduced with permission from: © Bettmann/CORBIS, p. 17; © Royalty-Free/CORBIS, p. 37.

Lerner Publications Company
A division of Lerner Publishing Group
241 First Avenue North
Minneapolis, MN 55401 U.S.A.

Website address: www.lernerbooks.com

Library of Congress Cataloging-in-Publication Data

Walker, Sally M.
    Light / by Sally M. Walker ; photographs by Andy King.
       p.   cm. — (Early bird energy)
    Includes index.
    ISBN-13: 978-0-8225-2925-5 (lib. bdg. : alk. paper)
    ISBN-10: 0-8225-2925-4 (lib. bdg. : alk. paper)
    1. Light—Juvenile literature. 2. Optics—Juvenile literature. I. King, Andy, ill. II. Title.
III. Series: Walker, Sally M. Early bird energy.
QC360.W346 2006
535—dc22   2005005651

Manufactured in the United States of America
1 2 3 4 5 6 – BP – 11 10 09 08 07 06

CONTENTS

BE A WORD DETECTIVE

Can you find these words as you read about light? Be a detective and try to figure out what they mean. You can turn to the glossary on page 46 for help.

absorbs	opaque	refraction
atoms	photon	translucent
energy	rays	transparent
matter	reflect	wavelength
molecule		

Playing games with light can be fun. But we also need light. Why is light important?

CHAPTER 1
WHAT IS LIGHT?

Have you ever tried to run away from your shadow? Have you ever made faces in front of a mirror? Without light you couldn't play these games.

Light is a form of energy. Light lets us see the world around us. Light also makes heat. The Sun gives off a lot of light and heat.

Light from the Sun makes life on Earth possible. Sunlight carries energy to plants and animals. Sunlight also carries heat to Earth. All water on our planet would freeze without the warmth from sunlight.

Most of our light energy comes from the Sun.

Fire makes light and heat too. A fire can be used to heat things or to make an area bright. Light energy from lightbulbs brightens buildings and neighborhoods. Some lasers use light energy to cut things.

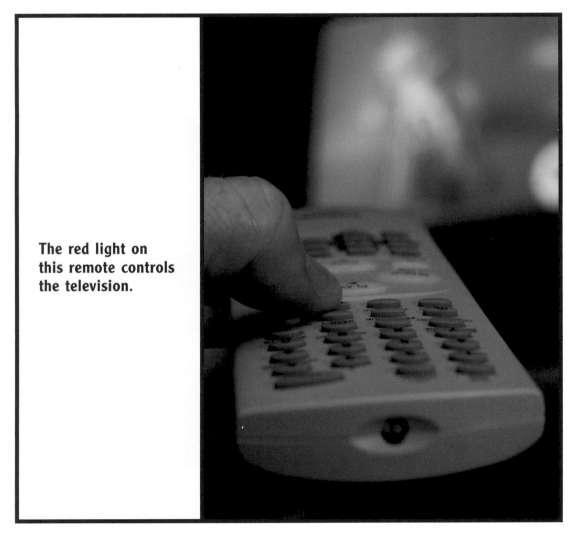

The red light on this remote controls the television.

Everything around you is made of atoms. Air, water, plants, furniture, and even your body are made of atoms.

Light is very important. But where does it come from? Light comes from tiny particles called atoms. Atoms are so small that billions of them would fit on the period at the end of this sentence.

Like you, atoms can get excited. An atom becomes excited when it gets extra energy. The particles inside an excited atom move faster. If the atom is excited enough, a small burst of light energy shoots out. This small burst of light energy is called a photon.

When you are excited, you probably jump up and down. When atoms from a light source get excited, they give off photons.

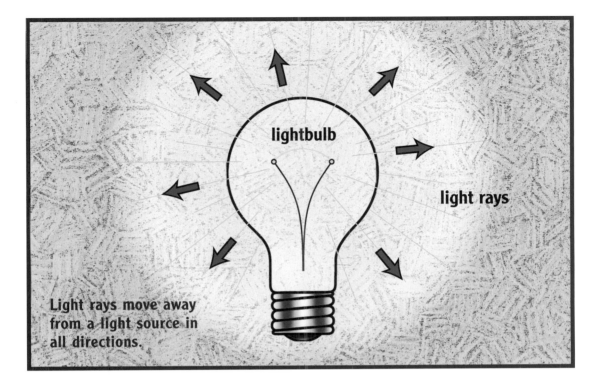

lightbulb

light rays

Light rays move away from a light source in all directions.

A photon streaks out from a light source. Excited atoms on the Sun shoot out photons all the time. The photons travel in rays. Rays are narrow beams of light. A laser's ray is a moving stream of photons. So are the rays of light from a flashlight.

Light rays move away from their source in all directions at once. That's why the light from one lightbulb can make an entire room brighter.

These kids are pretending to be atoms that are joined together. What do we call two or more atoms that have joined together?

CHAPTER 2
MATTER AND LIGHT

Everything around you is made of tiny atoms that have joined together. When two or more atoms join together, they make a molecule (MAHL-uh-kyool). A molecule is bigger than an atom. But it's still very small.

The rock is solid. The milk in the cup is liquid. What kind of matter is the cup?

Atoms and molecules make up matter. Matter is anything that can be weighed and takes up space.

Some matter is solid, like rocks and furniture. Some is liquid, like milk or water. Matter can also be a gas. The air we breathe is made up of gases. So what does matter have to do with light?

Usually we can't see light rays moving. But we can sometimes see light rays when they hit matter.

You can make light rays visible. You will need aluminum foil, a flashlight, a pin, a spoonful of flour, and a sheet of newspaper.

If you don't have flour, you can use baby powder or chalk dust.

Usually the light from a flashlight spreads out as it travels. But the light will travel as a narrow beam through a pinhole.

Spread the newspaper on the floor close to a wall. Cover the light-making end of your flashlight with aluminum foil. Use the pin to make a tiny hole in the center of the foil. Fill the spoon with flour and set it down on the newspaper.

Turn off the light in the room. Turn on the flashlight. Point it at the wall above the place where you put the newspaper. Can you see a spot of light on the wall?

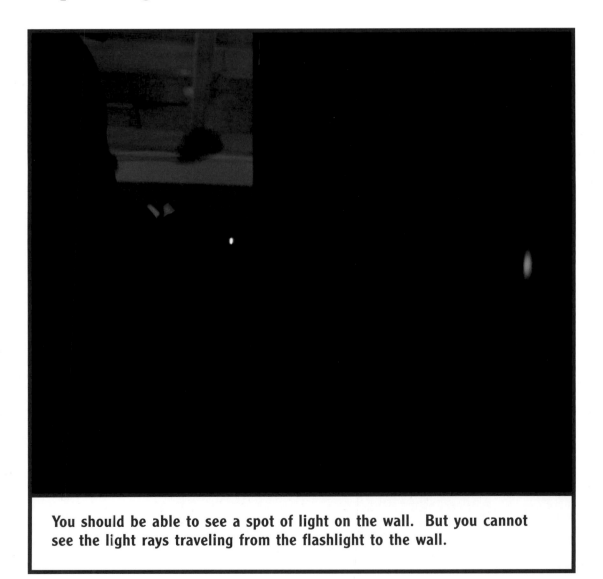

You should be able to see a spot of light on the wall. But you cannot see the light rays traveling from the flashlight to the wall.

Sometimes you can see moving light rays coming through a window on a sunny day. The rays are hitting dust and other matter in the air.

Pick up the spoonful of flour. Gently blow the flour toward the space between the flashlight and the spot of light. Can you see the light beam? Yes. The flashlight's rays travel in a straight line to the wall. You can see the rays now because they are hitting the flour, which is matter!

Light keeps traveling until it hits something. Then it bounces like a ball. What happens when light bounces off an object?

CHAPTER 3
LIGHT BOUNCES

We can see objects because light rays reflect off matter. Reflect means to bounce off. Light rays reflect off objects the same way a ball bounces off the ground. If an object reflects no light rays, we cannot see it. Prove it for yourself.

You can still see the objects in this room. Some light is reflecting off the objects to your eyes. Where is the light coming from?

You will need a room with a wall mirror and a flashlight. Make the room as dark as you can. If it is totally dark, you won't see the walls at all. That's because no light rays are reflecting off the walls. If you can see the walls dimly, some light is in the room. It is reflecting off the walls to your eyes.

Hold the flashlight in front of your chest. Turn it on and point it straight at an empty wall. The wall absorbs many of the light rays. Other rays are reflected back to your eyes. How do you know?

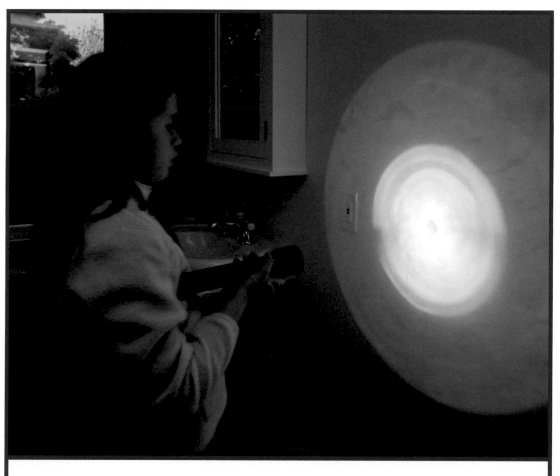

The flashlight's rays make a circle of light on the wall.

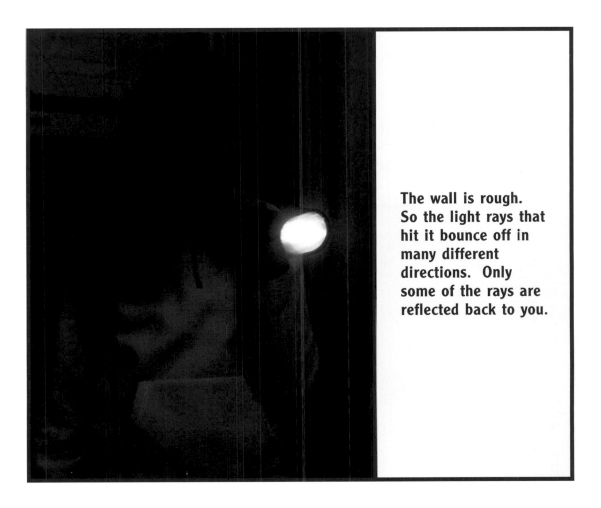

The wall is rough. So the light rays that hit it bounce off in many different directions. Only some of the rays are reflected back to you.

You know because you can see a circle of light where the rays reflect off the wall. The part of the wall that is not reflecting light stays dark.

Look down at your shirt. Is there a circle of light on it? No. Not enough rays are reflected to make a circle of light on your shirt.

Stand in front of a mirror. Hold the flashlight chest high again. This time, shine the light straight at the mirror. Look down at your shirt. Is there a circle of light? Yes. Why?

The mirror reflects enough light to make a circle of light on your shirt.

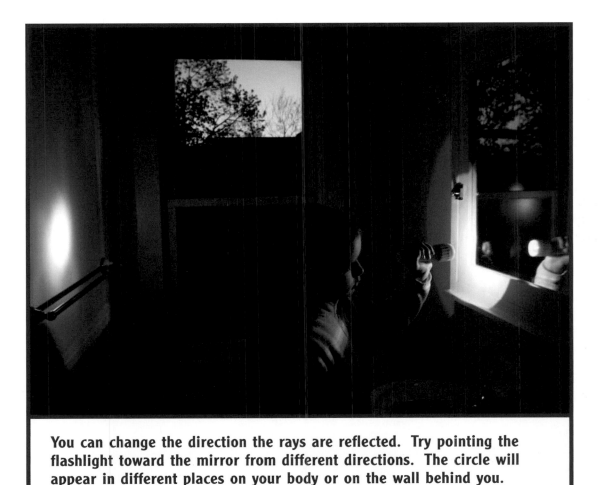

You can change the direction the rays are reflected. Try pointing the flashlight toward the mirror from different directions. The circle will appear in different places on your body or on the wall behind you.

The circle is there because the mirror reflects more light than the wall does. The mirror's surface is very smooth. Smooth surfaces reflect a lot of light. Most of the light rays striking the mirror are reflected back onto you.

23

Light shines easily through eyeglasses. What kind of matter are eyeglasses made of?

CHAPTER 4
LIGHT BENDS

Some light rays are not reflected. Instead, they pass through certain kinds of matter. Matter that light can pass through very easily is called transparent. Air is transparent. So are water and clear glass. Light rays shine easily through these materials.

Clouds are translucent when a little bit of sunlight shines through them.

Some matter only lets a few light rays shine through. That dims the light. Matter that some light can shine through is called translucent (trans-LOO-suhnt). Wax paper, milk, and thin cloth are translucent. Fog and wispy clouds are translucent too.

Some matter stops light rays completely. No light shines through them at all. This kind of matter is opaque (oh-PAYK). Bricks are opaque. So are aluminum foil and a thick block of wood. Your body is opaque. That's why you cast a shadow. A shadow is an area where light rays could not pass through.

Your body is opaque. It blocks light and creates a shadow.

Light travels very quickly. No matter how fast this car moves, the light from its headlights always travels faster.

Light rays travel very quickly. They travel much faster than you can blink your eyes. Light rays in outer space race at a speed of 186,282 miles per second. Light rays slow down when they enter Earth's air. Even so, a light ray from the Sun takes only eight minutes to reach Earth. A rocket would take many years to make the same trip.

Light rays can bend when they pass from one kind of transparent matter into another. Glass and water are both transparent.

Light rays can bend when they pass from one kind of matter into another. They bend because their speed changes. The bending of light rays is called refraction (rih-FRAK-shuhn). You can see light rays refract. You will need half a glass of water and a pencil or pen.

Put the pencil or pen into the glass. Let it lean against the side of the glass. Hold the glass so the top of the water is level with your eyes. Look at the pencil. It appears to be bent! Why?

This pen looks bent or broken when part of it is put into water. The part of the pen in the water also looks thicker.

The top of the pencil is in the air. Light rays reflected from this part of the pencil travel quickly. Light rays reflected from the bottom of the pencil travel through the water. Light rays travel more slowly in water than in air. When the light rays leave the water, they speed up.

REFRACTION

pencil

glass

water

Light rays slow down and bend when they enter water.

The pen looks straight when it is not in the water. The light rays bouncing off the pen are no longer refracted.

The rays bouncing off the bottom of the pencil are refracted as they speed up. The rays change direction and make the pencil seem bent. But if you pull the pencil out of the glass, you can see that it isn't bent.

Light might look white. But it is really made up of many colors. What kind of light is made up of every color?

CHAPTER 5
COLORS

Sunlight is called white light. But white light contains many different colors mixed together. It has all the colors in the rainbow! A simple experiment will help you prove this. You will need a shallow baking pan, water, a mirror, and a flashlight.

Fill the pan with water. Place it on a table or a counter near a wall. Lean the mirror against the edge of the pan farthest from the wall. Half of the mirror should be out of the water.

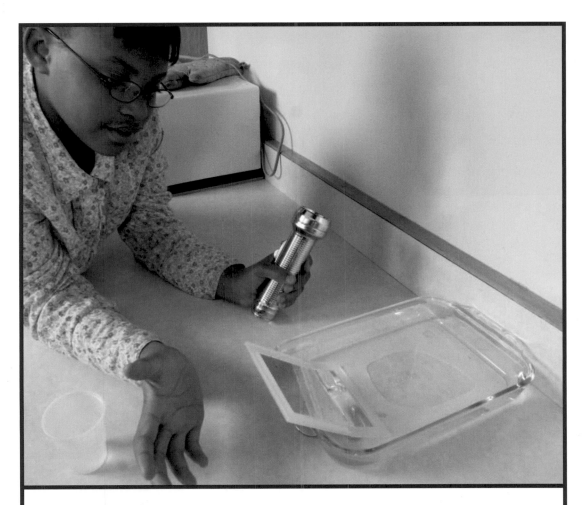

You might need to use some tape or put a weight in front of the mirror to hold it up.

Try lowering the mirror or moving the flashlight beam to get the rainbow to appear.

Shine the flashlight at the part of the mirror that is underwater. A rainbow will appear on the wall. You have split white light from the flashlight into many colors. What makes the colors appear?

Refraction makes the colors appear. Light acts like a wave. A wave is like a line that keeps moving up and down. The distance from one top point to the next top point is the wavelength.

Each color of light has a different wavelength. Blue light has a short wavelength. Red light has the longest wavelength. Each color of wave is refracted a different amount.

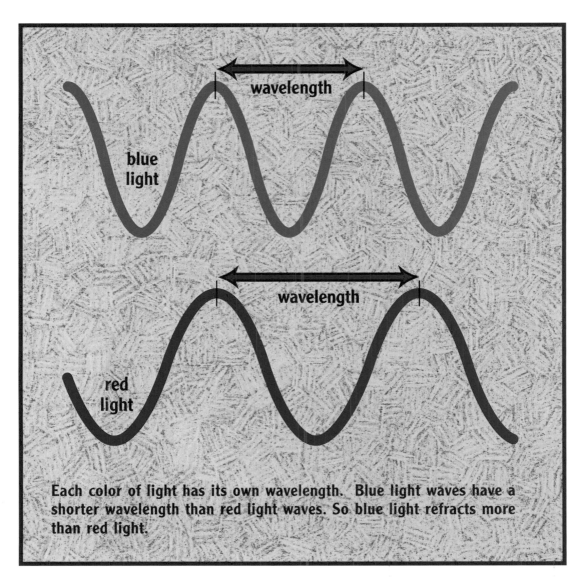

Each color of light has its own wavelength. Blue light waves have a shorter wavelength than red light waves. So blue light refracts more than red light.

When white light travels into the water, all the light waves are refracted. Each wavelength of light bends different amounts. This separates the white light into colors. Then the mirror reflects the waves to the wall. Our eyes see them as separate colors. We can see all the colors at once. We see a rainbow.

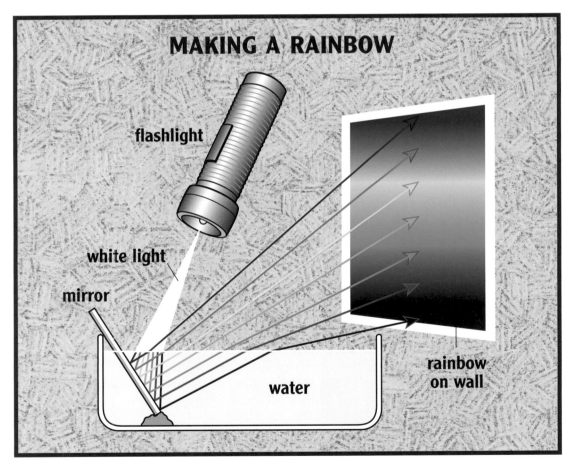

MAKING A RAINBOW

flashlight

white light

mirror

water

rainbow on wall

You see a red apple because all the colors except red are absorbed into the apple. Red light is reflected from the apple's skin.

Each object absorbs and reflects light waves differently. Because of this, objects have different colors. When an object reflects all light waves, it looks white.

When you see a red apple, you are seeing red light waves. The apple reflects red light waves. Light waves of other colors are absorbed into the apple.

When you are outside on a sunny day, you will feel warmer wearing a black shirt than wearing a white one. Do you know why?

CHAPTER 6
LIGHT AND HEAT

When an object absorbs all colors of light, it looks black. What happens to matter when light waves are absorbed?

For this experiment, use two glasses that are the same. If the glasses are different, the experiment won't work as well.

When light rays are absorbed, they make matter get hot. You can prove this with four pieces of paper, two white and two black. You'll also need tape, scissors, a thermometer, water, and two identical glasses.

Fill both glasses with cool water. Take the temperature of the water in each glass. Write down the temperatures. The water in both glasses should be about the same temperature.

Wrap one glass with white paper. Cover the top of the glass with a small piece of white paper. Do the same thing to the other glass using the black paper.

Put the glasses next to each other in direct sunlight. Wait 30 minutes. Take the temperature of the water in each glass. Have the temperatures changed?

Do this experiment in the middle of the day. The sun's rays shine brightest then.

The black paper absorbed all the light rays. The absorbed light heated the water in the glass. The temperature of the water increased.

The water in the glass covered with black paper is much warmer. Why? The black paper has absorbed all of the light rays. The absorbed rays heated the water.

The water in the glass covered with white paper hasn't heated up as much. Why? The white paper reflected most of the light rays. Fewer rays were absorbed. So the water didn't get as warm as the water in the other glass.

The white paper reflected most of the light rays.

You can see yourself in a mirror because of light. It is reflecting off the mirror to your eyes.

You have learned a lot about light. Sunlight warms Earth. Light brightens a dark place. It adds color to your life. The next time you look in a mirror, think of light. Without it, your reflection wouldn't be there!

A NOTE TO ADULTS
ON SHARING A BOOK

When you share a book with a child, you show that reading is important. To get the most out of the experience, read in a comfortable, quiet place. Turn off the television and limit other distractions, such as telephone calls. Be prepared to start slowly. Take turns reading parts of this book. Stop occasionally and discuss what you're reading. Talk about the photographs. If the child begins to lose interest, stop reading. When you pick up the book again, revisit the parts you have already read.

BE A VOCABULARY DETECTIVE

The word list on page 5 contains words that are important in understanding the topic of this book. Be word detectives and search for the words as you read the book together. Talk about what the words mean and how they are used in the sentence. Do any of these words have more than one meaning? You will find the words defined in a glossary on page 46.

WHAT ABOUT QUESTIONS?

Use questions to make sure the child understands the information in this book. Here are some suggestions:

What did this paragraph tell us? What does this picture show? What do you think we'll learn about next? What is a photon? What kind of matter can light pass through easily? What happens to matter when all light waves are absorbed? What was your favorite experiment? Why?

If the child has questions, don't hesitate to respond with questions of your own, such as: What do *you* think? Why? What is it that you don't know? If the child can't remember certain facts, turn to the index.

INTRODUCING THE INDEX

The index helps readers find information without searching through the whole book. Turn to the index on page 48. Choose an entry such as *shadows* and ask the child to use the index to find out how shadows are made. Repeat with as many entries as you like. Ask the child to point out the differences between an index and a glossary. (The index helps readers find information, while the glossary tells readers what words mean.)

LEARN MORE ABOUT LIGHT

BOOKS

Gibson, Gary. *Light and Color.* Brookfield, CT: Copper Beech Books, 1995. Experiment with sunlight, cameras, kaleidoscopes, moving pictures, dyes, and 3-D to learn the science behind light and color.

Murphy, Bryan. *Experiment with Light.* Princeton, NJ: Two-Can Publishing, 2001. Find out what light and color have to do with sundials, eyeglasses, optical illusions, television, and much more.

Nankivell-Aston, Sally, and Dorothy Jackson. *Science Experiments with Color.* New York: Franklin Watts, 2000. Explore the basics of color and light using paints, lights, material, sunlight, and plants.

Otto, Carolyn B. *Shadows.* New York: Scholastic Reference, 2001. Learn all about shadows in this easy-to-read book.

Tocci, Salvadore. *Experiments with Light.* New York: Children's Press, 2001. This book has many fun experiments that show how light and color work.

WEBSITES

Beakman & Jax's 50 ?'s
http://www.beakman.com/questionlist.html
Find answers to light-related questions about why the sky is blue, how magnifying glasses work, and why people have to wear eyeglasses.

Make a Splash with Color
http://www.thetech.org/exhibits_events/online/color/intro/
Colorful illustrations help explain where color comes from, what we can do with it, and how we see it.

ReviseWise Science: Physical Processes: Light
http://www.bbc.co.uk/schools/revisewise/science/physical/14_act.shtml
This interactive site includes fun activities, facts, and quizzes about light sources, reflection, and shadows.

GLOSSARY

absorbs: soaks up or takes in

atoms: very tiny particles that make up all things

energy: the ability to do work

matter: anything that can be weighed and takes up space. Matter can be a solid, a liquid, or a gas.

molecule (MAHL-uh-kyool): a group of atoms that are joined together

opaque (oh-PAYK): blocking light rays completely

photon: a tiny packet of light energy

rays: narrow beams of light

reflect: to bounce off of matter

refraction (rih-FRAK-shuhn): the bending of light as it passes from one material to another

translucent (trans-LOO-suhnt): letting some light pass through, but not all

transparent: lets all light pass through so that objects on the other side can be seen clearly

wavelength: the distance on a wave from one top point to the next top point

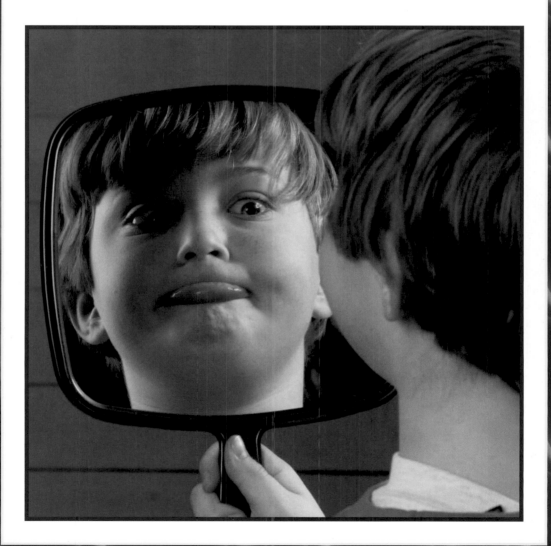

INDEX

Pages listed in **bold** type refer to photographs.

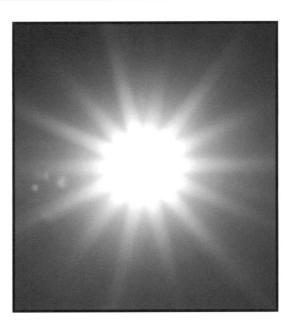